NASA's Preparation for its First Commercial Craft Moon Landing

Everything you need to know about the mission and its importance since Apollo 11

Alvin V. Grey

TABLE OF CONTENTS

INTRODUCTION

Final preparations are in place for the mission that aims to repeat the historic successes of the Apollo program by placing a U.S. lander on the moon, an undertaking not done in more than 50 years. The story of space travel is about to take another enormous turn as the world watches for the launch of Peregrine mission one, which has the name of the fastest animal on Earth.

Not merely another expedition in the history of lunar exploration, the Peregrine project marks a turning point in space exploration history where the domains of science and commerce collide. Peregrine is slated to blast into the sky on Monday at 7:18 am UK time, marking a new chapter in the ever-evolving history of space exploration. Although NASA provided sensors for the robotic lander, Pittsburgh-based Astrobotic is in charge of the mission, which is the first time a private corporation has attempted a soft landing on the moon or any other celestial body.

Chief Executive John Thornton of Astrobotic captures the general feeling of excitement, fear, and thrill associated with this endeavor. He acknowledges, "There's a lot riding here." The risk factor raises the stakes and complicates the unfolding narrative as the Peregrine lander is perched atop the United Launch Alliance's unproven Vulcan rocket. This mission's accomplishment not only represents a significant milestone for Astrobotic, but it also paves the way for more commercial lunar projects in the future.

Being the first mission under NASA's Commercial Lunar Payload Services (CLPS) project, Peregrine is a trailblazer in the field of lunar exploration. NASA works with commercial businesses under this creative plan, paying them to send scientific equipment to the moon. The variety of goals is demonstrated by the payloads carried by Peregrine: fifteen non-NASA payloads and five NASA payloads, including a shoebox-sized rover from Carnegie Mellon University that is expected to be the first American robot to go across the lunar surface.

But not every payload is strictly scientific; there are traces of humanity's varied pursuits tucked away in Peregrine's cargo. Alongside the technological tools are a copy of Wikipedia, a real coin containing one bitcoin, and DHL "moonboxes" filled with souvenirs that range from books and photos to a tiny piece of Mount Everest. Furthermore, cremated human bones and DNA—including that of Gene Roddenberry, the brilliant man who created Star Trek—find their way to the moon thanks to Elysium Space and Celestis.

These symbolic actions haven't, however, been without criticism. Native Americans expressed worries about the moon's holiness, especially in light of the inclusion of items that they believed to be sacrilegious. The commercial aspect of Peregrine is emphasized by Chris Culbert, the CLPS program manager at NASA's Johnson Space Center, who also emphasizes that NASA cannot control Astrobotic's payload selections. The lunar story is made more difficult by the conflict between business interests and scientific inquiry as well as cultural sensitivity.

The ancient lava flow that Peregrine is traveling to is called Sinus Viscositatis, sometimes known as the Bay of Stickiness—a term that comes from the peculiar nature of the lava deposits. If all goes as planned, Peregrine's equipment will detect the magnetic field, radiation levels, the exosphere—a very thin layer of gas around the moon—and surface and subsurface water ice. In addition to being scientific pursuits in and of itself, these readings are essential for risk reduction and utilizing the moon's resources for manned expeditions in the future.

The PITMS (Peregrine Ion-Trap Mass Spectrometer), a miniature mass spectrometer intended to examine the makeup of the lunar exosphere, is one of the main instruments on board Peregrine. The primary UK co-investigator on PITMS and senior research fellow at the Open University, Simeon Barber, admits that the mission carries a significant risk. PITMS tracks changes during the eight Earth days when Peregrine is in operation with the goal of deciphering the secrets of the lunar exosphere. The instrument's importance as a possible resource for future lunar activities is highlighted by its involvement in understanding lunar water circulation, particularly in light of temperature variations from 100C to -100C.

Under the CLPS plan, Peregrine is only the first of many landers—including Intuitive Machines—who will go on successive missions to the moon, heralding a lunar renaissance. While anticipation for these trips grows, several scientists express worries about the necessity of agreements to safeguard particular lunar sites that may be significant for future scientific installations.

NASA is returning to the moon as well as turning the pages of lunar exploration. The

Space Launch System (SLS) test flight on Nov. 16 marked the return of lunar goals after decades without a rocket strong enough for deep space exploration. The SLS, a 322-foot megarocket, marked the beginning of American astronauts' return to lunar exploration by launching the Orion spacecraft on its maiden empty test mission.

But the dedication to the moon is only the beginning of something bigger. NASA's Associate Administrator for Science, Thomas Zurbuchen, highlights that the Artemis mission is a first step toward the eventual goal of landing on Mars. The deep-space radiation, high temperatures, and lack of an atmosphere that characterize lunar exploration provide as testing grounds for technologies and knowledge that will be required for even more ambitious interplanetary missions.

Collaboration is required to realize this ambitious vision. NASA can no longer carry out these projects by itself. This new era of joint activities is reflected in the Artemis Accords, an international pact that sets norms for safe and cooperative space exploration. However, NASA is up against competition as it expands. China has made great progress with its military-run space program, establishing its own space station, Tiangong, and sending robotic missions to land on the moon. Many countries and corporate organizations are striving for a piece of the lunar narrative, and the moon has become one of their main focus points.

The moon, which was formerly a representation of human achievement and exploration, is about to take on new significance as a hub for space trade. Once believed to be an arid wasteland, the finding of water on the moon has changed perceptions and made the moon a possible center for scientific and commercial endeavors. Water that had frozen in shadowy craters could be extracted for drinking and separated into hydrogen and oxygen for use as rocket fuel and breathing gas. It appears that humans may use the moon as a space gas station in order to save money on the enormous expenses associated with bringing heavy fuel from Earth.

This is a paradigm shift that affects business as well as scientific research. Over the next 30 years, the sector for only mining lunar water is expected to be worth $206 billion. The moon is an eighth continent that is ready for economic development and industrialization, not just a place for astronauts to land. Astor Perkins managing partner Scott Amyx sees the moon as a launching pad for a variety of businesses, including energy and mining as well as real estate and tourism.

But as humanity explores this uncharted territory, concerns surface. The long-running argument over robotic vs. human exploration comes up again. Critics wonder why astronauts would be sent when robots could carry out the same duties without requiring air, water, or company. In response, Thomas Zurbuchen highlights how people are superior to machines in certain areas, such as situational awareness, decision-making, and flexibility. From this angle, the moon ceases to be merely a scientific test and instead serves as a nursery for supporting life off Earth and preparing for further explorations of space.

This book attempts to untangle the complex story of lunar exploration, from NASA's Artemis program and the emerging lunar economy to the Peregrine mission's daring attempt to land on the moon. It investigates how science, business, and human ambition converge against the backdrop of the stars. The reader is welcome to go across space with us as we explore the upcoming chapters, where the moon—once a far-off celestial body—becomes a physical frontier for human invention and discovery.

CHAPTER ONE

The Peregrine Mission

The Peregrine mission stands out as a beacon of human activity amid the wide emptiness of space, where the moon seems to be watching over us silently. This mission, named for the fastest animal on Earth, is about to make history by trying a soft landing on the moon, a feat that hasn't been done in more than 50 years.

With Peregrine mission one, the story of space travel takes an exciting turn as the countdown clock approaches zero and the entire world holds its breath.

History and Purpose

The Peregrine mission is more than just a random trip into space; it is the result of painstaking preparation, cutting-edge technology, and a shared ambition to push the limits of human achievement. Peregrine is fundamentally a robotic lander, a trailblazer in the rapidly changing field of commercial lunar exploration. Under the direction of Pittsburgh-based Astrobotic, a company established in 2007, this mission represents the company's entry into unexplored markets.

Peregrine's main goal is apparent: to land gently on the moon and become the first commercial spacecraft to accomplish this historic feat. With a mission success rate that has exceeded predictions, Astrobotic aims to make history. An additional level of intrigue is added by selecting Sinus Viscositatis, the Bay of Stickiness, as the mission's destination. This old lava flow promises a plethora of scientific discoveries with its strange structures that imply a unique consistency.

Details and Timetable of Launch

Peregrine's voyage starts in Florida at Cape Canaveral, a famous launch pad that has seen many space missions. Peregrine's launch, which is set for Monday at 7:18 a.m. UK time, is evidence of the accuracy and cooperation needed in space exploration. Launchpad vibrations echo as the rocket engines roar to life and Peregrine sets off on its celestial mission, leaving Earth's atmosphere behind in pursuit of the moon.

Strategically timed, the launch will allow the spacecraft to circle the Earth and then head toward its lunar rendezvous. This trajectory enables the best possible alignment and positioning, which are essential for the successful completion of a difficult mission. Peregrine's ascent is set against the exquisite dance of celestial bodies, emphasizing the careful calculations required to navigate the great reaches of space.

Difficulties and Risk

Every mission in the high-risk field of space exploration is extremely important, and Peregrine is no exception. The unproven Vulcan rocket Peregrine is perched on and the inherent uncertainty of the lunar landing itself present a plethora of threats. The general feeling is best expressed by Astrobotic CEO John Thornton, who combines exhilaration, thrills, and a hint of caution. "There's a lot riding here," he says, expressing the weight of uncertainty and anticipation.

There's an element of surprise added by the Vulcan rocket. Although Peregrine is the first payload to fly atop the unproven Vulcan, its manufacturer, United Launch Alliance, boasts a 100% mission success rate with its previous rockets. Every launch is, by definition, a step into the unknown and a monument to human ingenuity and the unwavering desire to explore boundaries beyond our home planet. This is the very nature of space exploration.

The difficulties get more intense as Peregrine rockets near the moon. Lunar landings have a history of both successes and failures, and they are infamously challenging. Modern technology and quick decision-making are needed to provide the precision needed for a gentle landing. The Peregrine mission is made more challenging by the fact that no private business has accomplished this accomplishment. It's an audacious endeavor, a measured risk that aims to exceed the limits of what commercial organizations can accomplish in the field of space exploration.

The Lunar Exploration Commercial Landscape

Peregrine is not just a scientific project; it represents a paradigm change in lunar exploration as well as a business venture into space. Under NASA's innovative Commercial Lunar Payload Services (CLPS) program, private businesses are hired to transport scientific equipment to the moon. This mission is part of this program. Because of its pioneering role in this mission, Astrobotic is positioned to lead the way in the developing field of commercial lunar activities.

A new dynamism in lunar exploration is introduced by the engagement of commercial firms. The moon is now used as a platform for commercial companies to display their capabilities, rather than just being the purview of government space agencies. Leading the way, Astrobotic paves the way for future joint ventures in space exploration between the public and commercial domains. The peregrine falcon becomes a symbol of a new era in which the moon is not only a place of scientific study but also a possible center of commercial activity.

The Peregrine expedition embodies both human curiosity and technological prowess as it rockets towards its lunar target. This chapter explores the complexities of the project, including its goals and beginnings as well as the hazards and difficulties that come with lunar exploration. With its blend of commercial and scientific goals, Peregrine becomes a microcosm of the changing story of space travel, in which the moon is not only a neighbor on Earth but a frontier waiting to be explored, ideas developed, and bold dreams come true.

CHAPTER TWO

The Commercial Lunar Payload Services (CLPS) Initiative

In the rapidly developing field of space exploration, where distinctions between government-sponsored programs and private sector ventures become increasingly hazy, the Commercial Lunar Payload Services (CLPS) program is an example of how lunar exploration is changing. The NASA-initiated program is a paradigm shift in lunar mission approaches that will allow private organizations to play a major role in furthering humankind's presence beyond Earth.

CLPS Genesis

The idea behind CLPS was the realization that working with the commercial sector may spur lunar exploration in ways never seen before.

When space organizations around the globe started thinking about going back to the moon in the early 21st century, NASA had an idea for a system that would take use of the efficiency and creativity found in the private sector. The CLPS initiative, a revolutionary method of lunar exploration that would advance scientific understanding and boost the growing space industry, was the realization of this ambition.

Goals and Structure

Fundamentally, CLPS aims to use private companies' resources to send research payloads to the moon. NASA chose a cooperative approach where private enterprises would compete for contracts to deliver research instruments, rovers, and other payloads to the moon rather than depending exclusively on conventional government-led missions. Beyond short-term scientific advantages, the initiative's goals are to establish a foundation for long-term lunar operations and human exploration.

The CLPS architecture is made to be adaptive and flexible. Under the initiative, NASA assigns task orders to organizations that specify mission needs and objectives. This competitive concept promotes creativity, economy, and a wide variety of submissions. Thorough assessments are conducted during the selection process to make sure the selected organizations fulfill the financial and technological requirements necessary for successful lunar missions.

Clinical Linkages Pioneered by Astrobotic and the Peregrine Mission

Leading the Peregrine mission is Pittsburgh-based Astrobotic, which stands out as a trailblazer in the CLPS program. Peregrine occupies a special place in the history of lunar exploration because it was the first mission operated under CLPS. The fact that Astrobotic was chosen for this momentous project is not only evidence of its competence but also establishes a standard for other businesses hoping to support lunar research.

The partnership between NASA and Astrobotic heralds a new age in which the public and commercial sectors work together more effectively. NASA assigns commercial companies such as Astrobotic the task of carrying out lunar missions due to its extensive experience and scientific goals. This alliance goes beyond a single mission; it establishes the foundation for an ongoing collaboration that may influence lunar exploration in the future.

Payloads and Variability of Goals

Missions such as Peregrine carry a wide range of scientific instruments and symbolic artifacts that fall under the purview of CLPS. NASA uses these missions to carry out research, collect information about the lunar surface, and clear the path for upcoming human exploration. The fact that Peregrine carries payloads from both NASA and non-NASA nations and commercial organizations demonstrates the initiative's inclusive and cooperative spirit.

The variety of payloads reflects the many goals of lunar exploration. Each payload adds a component to the overall picture of our understanding of the moon, whether it is through radiation levels, surface water ice research, or testing cutting edge technology. Furthermore, souvenirs from different sources, a copy of Wikipedia, and a physical coin with one bitcoin all serve as symbolic objects that provide a human touch to scientific pursuits by signifying humanity's shared cultural and adventurous legacy.

Difficulties and Debates

Although CLPS marks the beginning of a new age in collaboration, there are difficulties and disagreements with it. Discussions concerning the moral and cultural ramifications of commercial lunar operations have been triggered by the presence of symbolic objects, such as DNA and cremated remains, on lunar missions. Indigenous cultures have expressed concerns with the moon's sanctity, highlighting the importance of cultural sensitivity and awareness in these initiatives.

Furthermore, the difficulties involved with CLPS missions are exacerbated by the intrinsically dangerous nature of space travel. The high stakes are highlighted by the uncertainty surrounding lunar landings, the unproven rocket technologies, and the possibility of mission failure. In the enormous scale of lunar exploration, striking a balance between the goals of commercial entities, ethical considerations, and inherent hazards becomes a difficult act.

Prospects for the Future and Growth

CLPS prepares the way for a series of upcoming lunar missions as Peregrine sets out on its journey to the moon. The initiative's success depends not just on the individual missions but also on the total effect of many economic entities funding lunar exploration. With follow-up missions being planned and the CLPS framework providing a model for ongoing cooperation, the moon becomes the center of a wave of research and economic endeavors.

The CLPS mission is a component of NASA's larger Artemis program rather than an independent endeavor. CLPS acts as a forerunner to Artemis's goal of bringing people back to the moon through a sequence of flights that clear the path, test equipment, and collect vital data. Once a far-off place, the moon is now a dynamic frontier where science, business, and international cooperation come together to shape the course of human space travel.

This chapter explores the complexities of the Commercial Lunar Payload Services program, which is a revolutionary step forward for lunar exploration. With its collaborative spirit, CLPS not only transforms the interaction between public and private sectors, but it also leads humanity toward a future in which the moon serves as a platform for a variety of commercial, cultural, and scientific projects. As CLPS stands as a testament to the limitless possibilities that arise when humanity aims for the skies together, Peregrine is riding the crest of this revolutionary wave.

CHAPTER THREE
Scientific Payloads on Peregrine

With its daring objective of a soft landing on the moon, the Peregrine mission is more than just an exploration of space; it is a scientific journey with a cargo full of instruments meant to solve the mysteries surrounding Earth's closest friend. With a carefully selected payload, Peregrine is rocketing through space with the goal of exploring the lunar surface, determining its composition, and laying the groundwork for future human exploration.

Measuring the radiation levels on the moon is one of the main goals of Peregrine's research payloads. The lunar surface is vulnerable to solar and cosmic radiation since it lacks the shielding atmosphere that covers Earth. Comprehending these radiation levels is essential for organizing and ensuring the safety of next manned lunar missions. The instruments on Peregrine are set up to deliver important information that will guide the development of safety precautions for astronauts and their gear.

Peregrine's scientific investigation also focuses on the moon's surface and subsurface water ice. Finding water molecules on the moon, whether as ice on the surface or buried beneath the earth, is a huge deal. Water is an essential resource for upcoming moon missions since it may be used to make rocket fuel and possibly sustain personnel. The Peregrine research payloads has the potential to provide critical information that could influence the approach to utilizing lunar water resources.

Scientists are quite interested in the moon's magnetic field, which is a property that is present but less strong than Earth's. Among the equipment carried by Peregrine are tools intended to investigate the lunar magnetic field and provide insight into its properties and source. We can now better comprehend the moon's geological past and its long-term evolution thanks to this investigation of its magnetic environment.

The Peregrine Ion-Trap Mass Spectrometer (PITMS), a miniature mass spectrometer with a special purpose, is carried by Peregrine as it descends to the lunar surface. PITMS seeks to examine the makeup of the moon's extremely thin gaseous layer, known as the lunar exosphere. PITMS provides information on the composition, variability, and reactivity of the lunar exosphere to external stimuli by detecting molecules as they move across the lunar surface. Understanding the dynamic interactions taking place on the lunar surface is made possible by this knowledge, which is invaluable.

The mission's environmental research encompasses lunar surface exploration as well. Concerning the release and retention of water molecules throughout the lunar day and night cycles, researchers have put up some questions. With its accuracy, PITMS aims to understand the dynamics of lunar water molecules, including their release during the day and subsequent re-capture during the chilly nights. These findings have significant ramifications for comprehending lunar water circulation and evaluating lunar water's potential as a resource for upcoming lunar missions.

The mission has high stakes, as expressed by Simeon Barber, a senior research researcher at the Open University and the principal UK co-investigator on PITMS. The rover crew has even been given an unusual task: doing a doughnut motion to create a gas bubble on the moon's surface. These drastic efforts demonstrate the mission's dedication to gathering the most precise and thorough data possible.

With their varied goals, Peregrine's scientific payloads represent the spirit of inquiry and discovery that characterizes humankind's space travel. The cooperation between NASA, Astrobotic, and other scientific organizations highlights the shared goal of expanding human knowledge. As the Peregrine's instruments begin to function, they represent the coming together of scientific inquiry and technical innovation, solving the mystery of the moon and opening the door to the next phase of space exploration.

CHAPTER FOUR

The Controversies

Disputations are an unavoidable thread in the vast fabric of space exploration, where ethical concerns, cultural sensitivity, and scientific curiosity all mesh together. As it advances humanity toward a new era of lunar exploration, the Peregrine mission is embroiled in a tangle of discussions that call into question how space operations intersect with cultural heritage, morality, and the sanctity of celestial bodies.

A prominent controversy pertaining to the Peregrine expedition is the carrying on board of symbolic objects, DNA samples, and cremated human remains. A copy of Wikipedia, a physical coin loaded with one bitcoin, and DHL "moonboxes" filled with mementos ranging from novels and photos to a small chunk of Mount Everest are among the fragments of earthly existence carried by the spacecraft as it sets out for the moon. But what's really sparked a moving discussion is the presence of DNA and cremated human bones, some of which belonged to well-known people like Star Trek creator Gene Roddenberry.

The president of the Navajo Nation, Buu Nygren, expressed concerns over the act of leaving such artifacts on the moon in a letter to NASA, highlighting the sacred significance of the lunar surface for many Indigenous nations. Nygren's opinions are in line with a wider conversation over the moral ramifications of using the moon—or any other celestial body—as a holding place for human remains. It begs the question of whether this kind of action constitutes desecration or is it a valid way for mankind to show its connection to the universe.

In response to these worries, Chris Culbert, the CLPS program manager at NASA's Johnson Space Center, emphasized that Peregrine is a commercial mission and that NASA does not impose restrictions on Astrobotic's ability to fly. The ethical issues get more complicated when one distinguishes between a commercial endeavor and a mission directed by the government. Although private entities operating in space are subject to less regulatory restraints, government agencies may be required to follow specific protocols and international agreements.

There are more debates about Peregrine than just the presence of human and symbolic remains. There are concerns regarding how mankind should strike a careful balance between scientific discovery and cultural respect in light of the growing discourse surrounding the commercialization of space and lunar operations. Once a representation of mystery and awe, the moon is increasingly being used as a stage for business ventures, which may change how different cultures view it.

The discussion takes on a more futuristic bent with the addition of an actual coin loaded with one bitcoin. A byproduct of technological advancement on Earth, cryptocurrency makes its way into the cosmos on board Peregrine. The symbolic act of putting a digital money on the moon raises questions about how technology, finance, and space exploration are increasingly interacting. Humanity is facing novel and unprecedented challenges regarding the representation of value outside our planet as it expands its reach into space.

Understanding the thin line that separates scientific advancement and cultural sensitivity becomes crucial when addressing these disputes. Although scientific pursuits have traditionally dominated space exploration, the involvement of commercial organizations creates a dynamic in which exploration and profit-making are intertwined. This paradigm change raises questions about the obligations of individuals who venture into the cosmic frontier and calls for a reevaluation of the ethical frameworks guiding space activities.

International agreements and norms governing the use of celestial bodies are vital, as seen by the issue over Peregrine. Discussions concerning safeguarding heavenly bodies' holiness and safeguarding areas of special importance are becoming more important as the moon turns into a hub for business and scientific endeavors. Concerns over the possibility of unrestricted exploitation and manipulation of lunar landscapes are raised by the lack of explicit laws.

Even though there have always been debates over lunar missions, Peregrine offers a moving case study for understanding the complex world of space ethics. As the world enters a new phase of lunar exploration marked by partnerships between public and private sectors, the difficulties of striking a balance between financial interests, cultural sensitivity, and scientific research get more complicated.

One main question that comes out of all of these debates is: How can humankind explore the cosmos while maintaining the morals and values that make up our shared identity? With all of its financial and scientific ramifications, Peregrine becomes a focal point for answering this question and sparking a larger discussion about the direction space exploration will go in the future and the moral issues that must be resolved before venturing into the unknown.

CHAPTER FIVE

The Lunar Destination: Sinus Viscositatis

The ancient lava flow known as Sinus Viscositatis, or the Bay of Stickiness, is the endpoint of Peregrine's meteoric flight across the vastness of space, rather than some random place on the lunar surface. This chapter examines the significance of choosing this mysterious place for the Peregrine mission as well as the mysteries surrounding this lunar destination.

The Sinus Viscositatis Name

Sinus Viscositatis, the name alone, piques interest and conjures up a clear picture of a bay with an odd quality: stickiness. This odd name comes from the formations that have been found, which indicate that the lava that was formerly in this area had a unique consistency. Sinus Viscositatis challenges the common perception of the moon as a barren and arid expanse by suggesting that lunar geology has previously experimented with surprising textures.

Sinus Viscositatis Scientific Objectives

The reason for Peregrine's trip to Sinus Viscositatis is not random; rather, it is a purposeful decision motivated by scientific goals. Uncovered beneath the old lava flow is a geological narrative. The Peregrine spacecraft's equipment are ready to perform a symphony of measurements that will reveal the makeup, features, and background of this fascinating lunar location.

The measurement of radiation levels is one of the main goals. Because of its distinct geological characteristics, Sinus Viscositatis offers a chance to learn more about how radiation interacts with the lunar surface under particular topographical circumstances. This information advances our knowledge of the effects of cosmic radiation on celestial bodies and is essential for the security of upcoming lunar missions.

At Sinus Viscositatis, surface and subsurface water ice—a major Peregrine mission focus—comes into vivid relief. The old lava flow may have contained water ice, which could provide insights into the moon's hydrological past. Through an examination of the distribution and properties of water ice in this area, researchers hope to reconstruct the history of lunar water resources.

Another aspect of the moon that Peregrine's equipment will study at Sinus Viscositatis is its magnetic field, a modest but important trait. By illuminating the processes that created the moon's ancient lava flows and geological formations, our understanding of the magnetic characteristics of this region helps us understand the moon's geological evolution.

Sinus Viscositatis PITMS

The equipment known as the Peregrine Ion-Trap Mass Spectrometer (PITMS) lies at the heart of Sinus Viscositatis' scientific investigation. With its accuracy and analytical capabilities, this small mass spectrometer is assigned the mission of examining the makeup of the lunar exosphere at this particular site. PITMS takes a momentary picture of the distinct exospheric conditions in Sinus

Viscositatis as it sniffs molecules moving across the lunar surface.

In addition to revealing the makeup of the lunar exosphere, researchers hope that the PITMS observations will shed light on how this delicate layer of gas varies during the eight Earth days that Peregrine spends operating on the moon. The observations are further complicated by temperature variations caused by the moon's day and night cycles, which range from searing highs to bone-chilling lows.

The scientific world is waiting for the data that will come from Sinus Viscositatis as the rover crew plans unorthodox maneuvers, like a doughnut to release gases on the lunar surface. It is possible that the findings at this old lava flow will contribute to our knowledge of lunar geology, resource distribution, and environmental dynamics.

Lunar Prospects and Water as a Resource

Beyond the goals of science, Sinus Viscositatis represents an area of inquiry into the moon's potential as a resource for upcoming lunar expeditions. The finding of water ice in this area may have significant ramifications for long-term human habitation on the moon. Water is a valuable resource for space travel since it may be used to power rockets, hydrate people, and provide the structural support for lunar bases.

Regions like Sinus Viscositatis become vital checkpoints for evaluating the viability of human presence when humanity considers a return to the moon and the development of lunar outposts. With its distinctive geological characteristics, the old lava flow provides insight into the moon's past and may have some bearing on how humans may fare in space in the future.

By exploring Sinus Viscositatis as a lunar destination, this chapter highlights the careful planning and scientific purpose of the Peregrine mission. The mysteries of Sinus Viscositatis are revealed as the spacecraft descends onto the ancient lava flow, providing a fascinating window into the moon's geological past. With their ability to measure, probe, and analyze, Peregrine's scientific instruments will be able to reveal the mysteries of the Bay of Stickiness and add to the growing body of knowledge on lunar exploration.

CHAPTER SIX

CLPS's Future

NASA's Commercial Lunar Payload Services (CLPS) program represents a revolution in lunar exploration. In addition to marking the achievement of painstaking preparation and technological ingenuity, Peregrine's historic mission also serves as a predictor of CLPS's future course.

Compared to earlier space exploration models, which emphasized government agencies, CLPS is a change. Private businesses are essential in this new period, as they are turning the lunar surface into a vibrant marketplace. This change is more than just a sign; it's a calculated move by NASA to use the resources and expertise of the commercial sector to further its goals for lunar exploration.

As the first flight under CLPS, the Peregrine mission paves the way for a future wave of commercial lunar landers. The Pittsburgh-based company Astrobotic, which is in charge of the Peregrine mission, is a prime example of how well NASA and private industry work together. The accomplishments of Peregrine set the stage for the other CLPS missions, all of which add to the expanding corpus of lunar knowledge.

NASA's Commercial Partnership Investment

NASA's dedication to promoting partnerships with commercial partners is the foundation of CLPS. The program is an example of a planned approach to space exploration in which commercial enterprises are given government financing to send scientific payloads to the moon. This strategy fosters a thriving lunar economy with a diverse participant base while simultaneously quickening the pace of lunar missions.

NASA's financial support of businesses such as Intuitive Machines and Astrobotic, both of which are carrying out their maiden lunar missions under CLPS contracts, point to a larger trend. NASA's current budget commitment for CLPS through 2028 is $2.6 billion, demonstrating the agency's ongoing commitment to supporting a healthy lunar exploration ecosystem.

Increasing the Moon's Frontier
CLPS opens up new frontiers for lunar exploration. The historical Apollo landing sites are no longer the only thing on people's minds.

A wide range of lunar locations are made accessible by CLPS missions, each with its own distinct geological features and scientific possibilities. These missions aim to provide a thorough understanding of our planetary neighbor by solving the mysteries surrounding the craters, lava plains, and poles of the moon.

The variety of payloads that CLPS missions are capable of delivering is a reflection of the complexity of lunar exploration. These lunar landers house scientific equipment, rovers, and even symbolic objects like cremated human remains and a copy of Wikipedia. With CLPS, the moon becomes a platform for research and development, a laboratory for technology, and even a storehouse of human history.

Opportunities and Difficulties

There will be difficulties for CLPS in the future. Careful navigation is necessary in the complex dance between government agencies, private businesses, and multinational collaborations. Agility and creativity are brought about by the collaborative model, but it also calls for strong regulatory frameworks to guarantee moral and responsible behavior in space operations.

Concerns regarding the preservation of lunar sites are becoming more and more important as CLPS missions increase in frequency and variety. Some scientists, who are thinking ahead to future lunar colonies and observatories, support agreements to protect areas of special significance. Debates over the future of lunar activities center on the fine line that separates exploration from conservation.

Moon Mining: An Outlook for the Future

Though scientific research is still at the core of CLPS, the project also invites dialogue over the use of lunar resources. Water has been found in the polar areas of the moon, and it is vital to the survival of human habitation in the future. Originally considered a destination for scientific research, the moon is today considered a possible supply of vital resources for space missions.

As advocated by space resource utilization advocates, mining for lunar water could lead to the development of a lunar economy. The idea of taking water and using it for fuel production, oxygen synthesis, and drinking poses a paradigm shift in how we think about space operations. The moon seems as a celestial oasis that might sustain a long-term human presence and act as a springboard for further space research because of its enormous water ice reserves.

More than just a single mission, Peregrine's historic descent onto the lunar surface heralds the start of a new era in lunar exploration. With its emphasis on cooperation and business, CLPS is paving the way for a time when the moon is a hive of scientific research, technological advancement, and economic activity.

Collaboration between public and private sectors, ethical use of lunar resources, and preservation of lunar landscapes are critical to the success of CLPS. The future of CLPS goes beyond a single mission; it encompasses a long-term presence on the moon that will create the groundwork for a human return and open the door for ambitious missions that will come after.

The CLPS program is evidence of our shared goals as a species, as we turn our attention back to the moon. The lunar frontier, formerly only visited by a few, is now calling explorers from a wide range of sectors and backgrounds. The lunar canvas that serves as the backdrop for CLPS's future invites us to envision a time when the moon serves as more than just a destination—rather, it serves as a dynamic platform for exploring the secrets of the cosmos.

CHAPTER SEVEN

NASA's Moon Return

NASA's mission to the moon is more than just a trip back to a well-known planetary neighbor; it represents a significant revival of lunar exploration and a daring step towards more ambitious interplanetary goals. NASA's initiative, the Artemis program, embodies the agency's dedication to bringing people back to the moon and setting the stage for upcoming space exploration projects.

The Space Launch System and Artemis I

With the successful launch of the Orion spacecraft and the Space Launch System (SLS), the Artemis program made significant progress. The 322-foot megarocket broke apart in the sky on November 16 in a momentous event reminiscent of the Apollo era's Saturn V launches. Despite not carrying a crew, Artemis I was a significant step toward building the infrastructure required for crewed lunar missions.

An essential part of Artemis is the Space Launch System, a massive rocket intended for deep-space missions. Its capabilities are significantly beyond those of earlier launch vehicles, giving spacecraft the necessary push to get past low Earth orbit. The Artemis program's crewed missions were made possible by Artemis I, which showcased the capabilities of this revolutionary launch technology.

Artemis's Role for Orion

The crewed spacecraft at the center of Artemis, Orion, was also essential in Artemis I. Although it was not manned for this trip, Orion is scheduled to transport humans on future lunar expeditions. Orion's cutting-edge safety systems and design guarantee the astronauts' safety on their missions to and from the moon.

The Crewed Lunar Orbit and Artemis II

The next mission in the series, Artemis II, is scheduled to be the first Space Launch System and Orion crewed flight. In order to try a lunar landing, astronauts must first orbit the moon on this mission. Essential system and process testing is made possible by the moon orbit, guaranteeing that every detail is well tuned for the upcoming crewed lunar landing in Artemis III.

Artemis III and the Moon Return

Artemis III, where men will step foot on the lunar surface for the first time since the Apollo missions, is the apex of NASA's return to the moon. This mission, which might begin as early as 2024, is momentous because it intends to make history by sending the first woman and person of color to walk on the moon. Artemis III's extended lunar stay will enable vital research and resource usage trials, as well as establish the foundation for a long-term human presence on the moon.

Mars on the Horizon, Past the Moon

Although lunar exploration is the primary focus at the moment, NASA's commitment to going back to the moon should be viewed in the larger context of going beyond. NASA's assistant administrator for science, Thomas Zurbuchen, highlights that the mission to the moon is closely related to the mission to Mars. The intricacies and difficulties of a Mars mission exceed those of the Apollo period, necessitating international cooperation.

The Artemis Agreements and Global Cooperation

NASA understands that the mission to Mars is a team effort that cuts beyond national borders. An international pact known as the Artemis Accords sets guidelines for cooperative and safe space exploration. Participating nations in the Artemis program pledge to uphold values that guarantee openness, compatibility, and the peaceful use of space. The agreements highlight how crucial collaboration is to furthering humanity's cosmic exploration.

Opportunities and Difficulties

There are opportunities and challenges associated with NASA's return to the moon. Estimates indicate that the Artemis program will require significant financial outlays; by the time the first crewed mission lands on the moon, the budget might be as high as $93 billion.

CHAPTER EIGHT

Lunar Economy and Resources

The moon, which was formerly a far-off and enigmatic celestial entity, is becoming more and more of a viable frontier for resource development and commercial endeavors. The strategic application of lunar resources is becoming more and more important as human interest in lunar exploration grows. In Chapter 9, the intriguing potential of lunar resources and the imagined lunar economy are examined, along with the opportunities that exist within the lunar surface's aridity.

The Moon's Resources

Finding significant resources on the moon is one of the main reasons for the resurgence of interest in lunar exploration. It has been determined that the permanently darkened areas close to the lunar poles contain water in the form of ice. The fact that water can be transformed into hydrogen and oxygen, two elements needed to make rocket fuel, makes this discovery revolutionary. In addition to lowering the price of deep space travel, the possibility of using lunar water as fuel creates opportunities for longer journeys and refueling facilities outside of Earth.

The moon is abundant in several metals and minerals besides water. The moon's surface is covered in a loose, broken layer called regolith, which has substantial concentrations of iron, titanium, aluminum, and other elements. These metals could be used as building and manufacturing raw materials to support infrastructure on the moon in the future. The moon, which was formerly thought to be a desolate place, has the potential to develop into a celestial body rich in resources.

Resource Usage and Lunar Mining

Science fiction is no longer the exclusive domain for the idea of lunar mining. In order to satisfy the demands of future space travel, forward-thinking members of the space industry are actively investigating the potential of collecting resources from the moon. The process of lunar mining entails taking valuable resources from the lunar surface, and a number of methods are now under consideration.

Using robotic mining equipment to gather and process lunar regolith on-site is one suggested approach. The goal of this procedure, called in-situ resource utilization (ISRU), is to use the moon's resources for a variety of objectives. Engineers and scientists envisage metal and mineral extraction operations on the moon that will supply the building blocks needed to build spacecraft, housing, and other infrastructure.

The Shape of the Lunar Economy

The idea of a lunar economy starts to take shape as access to lunar resources increases. The lunar economy, which is motivated by the extraction and use of the moon's resources, aims to create a robust and sustainable ecosystem of economic activity there. This economic model turns the moon into a travel destination that can be sustained on a commercial basis, going beyond scientific study.

The construction of lunar bases and research stations is a crucial aspect of the lunar economy. These establishments would function as centers for technology advancement, resource exploitation, and scientific research. A wider spectrum of economic activity is made possible by the infrastructure required for a long-term human stay on the moon.

Lunar Activities and Space Tourism

In the imagined lunar economy, lunar sites are not immune to the appeal of space tourism. Private businesses have shown interest in providing lunar tourism experiences, enabling the general public to visit the moon and observe Earthrise from its surface. Firsthand lunar experiences present a new frontier for the tourist sector, with lunar resorts and exploration packages starting to appear on the commercial space scene.

Furthermore, moon activities go beyond travel. Businesses are looking into the possibility of carrying out studies, trials, and even movie shoots on the moon. The lunar environment's peculiarities provide a fresh setting for a range of creative and scientific pursuits. Once a far-off dream, the moon is now a blank canvas for human imagination and exploration.

Difficulties and Ethical Issues

A lunar economy has exciting potential, but it also presents a number of difficulties and moral dilemmas. The delicate lunar environment needs to be managed carefully to avoid causing ecological harm. Scientists and decision-makers are urging prudent lunar resource exploitation methods that prioritize environmental stewardship and sustainability.

The distribution of lunar resources also calls into issue international collaboration and equity. Frameworks that are inclusive and transparent are necessary to guarantee equitable access to lunar resources and opportunities for both private companies and national governments launching lunar missions. To promote a common understanding of humanity's future in space, the lunar economy ought to be based on the ideas of cooperation and mutual advantage.

Final Thought: A Moon Renaissance:

In the last chapter, a picture of a lunar renaissance is presented, in which the moon is transformed from a heavenly neighbor to a bustling center of human exploration and economic activity. By incorporating lunar resources into our economic pursuits, we can explore new avenues for creativity, teamwork, and human expansion beyond Earth. In the future, the moon will not only be a destination but also a vibrant ecosystem that adds to the larger picture of space exploration, as the lunar economy begins to take shape.

CONCLUSION

As we come to the close of this investigation into the fields of lunar missions, business ventures, and the aspirations of establishing a permanent human presence on the moon, we are at the cusp of a new age in space exploration. The preceding chapters have elucidated the intricacies, achievements, and disputes related to lunar pursuits. We are now, as we consider our shared journey through the pages of this book, at the beginning of a lunar expedition that is full of opportunities and difficulties.

Space exploration is at a turning point thanks to the Peregrine project and its bold goal of achieving a soft landing on the moon. A new paradigm is being ushered in with the combination of government-backed programs such as NASA's Artemis program with the rise of private companies under the Commercial Lunar Payload Services (CLPS) initiative. Once a far-off and unexplored planetary neighbor, the moon is today the focus of a wide range of commercial, scientific, and exploration endeavors.

Debates on resource management and ethical quandaries concerning cultural artifacts are just two of the controversies that are intertwined with lunar exploration, underscoring the complex nature of space travel. Maintaining a delicate balance while extracting valuable resources from the moon's surface is crucial, since it involves not only respecting the lunar environment but also the cultural and ethical considerations associated with our lunar aspirations.

The Peregrine mission's selected location, Sinus Viscositatis, becomes a metaphorical platform for scientific investigation. Instruments like the Peregrine ion-trap mass spectrometer (PITMS) will be used to detect radiation levels, examine water ice, and examine the lunar exosphere on this ancient lava flow. These scientific payloads' discoveries open the door for more human missions while lowering risks and utilizing the moon's resources.

Going ahead, the potential for a lunar economy and the use of lunar resources creates opportunities for technical advancement, economic expansion, and space travel. The moon, which was formerly a desolate region, is now a bustling center where governments, corporate enterprises, and scientists come together to pursue space exploration.

Thinking back on the difficulties and possibilities described in the previous chapters, we understand that the lunar odyssey is not an isolated undertaking. It is a shared experience that cuts over boundaries, philosophies, and academic fields. The moon invites humanity to venture beyond the boundaries of our home planet and discover the immense cosmic tapestry that lies ahead of us with its silent craters and ancient landscapes.

We tackle the difficulties of venturing into unknown waters in the chapters devoted to obstacles, opportunities, and the lunar economy. As we negotiate the lunar problems, technological barriers, moral dilemmas, and the necessity of international collaboration become guiding beacons, making sure that our attempts leave a good impact on the celestial body that has enthralled humanity for ages.

We are standing on the edge of a future in which the moon is more than just a destination—rather, it will be a monument to human creativity, tenacity, and curiosity—as we bid adieu to this investigation of lunar missions and the developing story of human interaction with the moon. The lunar voyage goes on, urging us to take an active role in the continuing space exploration narrative. May the knowledge gained, the obstacles surmounted, and the hopes sparked by reading these pages inspire us to work toward a time when the moon serves as more than just a destination—rather, it is a beacon that points us in the direction of the countless opportunities the universe has in store for us.